Heaven's Blueprints:

Proving YHWH's Creation & Designs on the Molecular Level

Introduction by: Olivia Leverenz

Written by: Destiny Hancock

We dedicate this book to YHWH for His divine appointment of our friendship, to our Lord and Savior Jesus Christ, and to the Holy Spirit for the Father's knowledge and wisdom that was poured out upon us while writing this book.

I dedicate this book to my parents, Andrew and Jenifer, for never telling me that I can't achieve my dreams and for always pushing me to be my best in following the Lord. To my loving husband, A.J., for your encouragement that fuels the Holy Spirit fire within me, even when I didn't know it was there. - Destiny

---

I dedicate this book to my mom, Julie, for the selfless sacrifices that she made while raising me and for loving me unconditionally and to Jake, who's the best brother that a girl could ever ask for. - Olivia

4

DESTINY

I was baptized by the Holy Spirit at 12 years old, and I began to prophesy in my local church. The head Pastor spoke to my mom in front of me that I had the anointing of Amy Semple McPherson. From the ages of 12 to 15, my family traveled and taught at many churches with a prophet named Tim Dodge, may he Rest in Peace. Little did I know, he was training me up in the way I would go. At 13 years old, the Holy Spirit revealed to me in a dream the call on my life, and to say I was freaked out would be an understatement — though, this book isn't about my testimony. That's for another time. The week before prophet Tim passed away, my mom had taken me to visit with him. At this visit, he began to pour encouragement into my life and said that I was anointed the same way that Deborah, Priscilla, and Peter had been anointed. Before we left his house, He prayed over me and passed the mantle of a prophet onto me. He passed away the following week and went home to the Father in 2010.

In 2014 I was at a revival in my hometown. This revival lasted three days, and on the last night, there was a passing of the mantle from one generation to the next, in the same kind of way that the prophet Elijah passed his mantle on to Elisha. I was raised in a family full of medical professionals, and by age of 5, I was trained to name all the bones in the human body on demand. Thus beginning my love for science. Culture teaches us that either science is true or God is true, but it can never be both. Why is that too hard to believe? God is the creator of ALL things, which means He created our understanding of science. Everything I studied left me with an insatiable desire to bring it back to God's Word. I never understood the reason why until I remembered the words that my mom constantly spoke to me, "Everything comes back to one truth." YHWH is that One Truth!

As the mantle was being passed to everyone in the room, the Holy Spirit began to minister to me. At the time, I was working on my Bachelor's degree in Biology and I was also working in the medical field. The Holy Spirit said, "Get a red notebook, get a red pen and begin to write the book." Confused with what was just spoken, I asked the Lord what book I was supposed to write. YHWH responded, "My people are confused. They search for proof of Me and they reject Me because they can't see the evidence that I've already provided for them. You see it. You know it. You've studied it and have brought it back to Me for years. I wrote My name on their hearts and now I want you to make them see the truth." I was humbled, yet terrified. I prayed again, "Why the red notebook and pen?" and with a laugh from the Holy Spirit I heard this response back to me, "Because My words are always written in red."

Engulfed in fire with the Holy Spirit from that night, I bought the notebook, red pen and began to write. Then that notebook sat on my desk untouched for the next six and a half years. I didn't feel worthy or qualified enough to accomplish this task that was assigned to me. Surely YHWH must have gotten it wrong by telling me what He called me to do or I had deluded myself into believing that I could possibly do it. Over the course of this time, I struggled like the

Israelites in the wilderness. Every single school opportunity failed. Every single job I had failed. The Lord blessed me through my shortcomings that would lead me to my husband and to our two miracle children we had together. Despite all of this, any career that I pursued failed and, unfortunately, I had to take emergency leave from school. I would cry out to YHWH in desperation for an answer, and the only response He gave me back to me was, "I already gave you the assignment."

    Fast forward to the end of 2020, two occurrences happened that woke me up to finish this book, when I only had 3 chapters written down. The first event was a situation of the parable, the buried talent. YHWH allowed me to witness other people being given fragments of the wisdom that He had also entrusted me with. He told me that He would remove His anointing off of my life and that it would not return. The second event was that He took me to hell. The Lord showed me the place that was prepared for me due to my disobedience and for rejecting His voice over the assignment that He gave me. As I was being shown all of this, all I could think about was the complete separation I felt from Him in that torturous moment. The torment all around me. The despair. The screams and when remembered, the smell of rotting flesh that still burns the inside of my nose. My eyes were finally opened. I very quickly understood in my spirit that I had placed myself here due to my own choice. It was overwhelming, convicting and a Matthew 7:21 reality check. After this, I began to write and eventually I finished the book. Every single word that was written down, the Holy Spirit wrote for me. It was my hand that He used, but it was God's words and His wisdom.

Throughout this time of wrestling with God, I had many faith-filled friends that would come into my life, but not long after they arrived, they would immediately exit stage left. I became depressed and anxious, believing that I was not meant to have friends, at least not the type of friend that I had been praying for. It's not that my husband isn't my best friend, but more so that I had desired to have an on fire, Holy Spirit filled woman of God in my life. A friend and a sister in Christ that would push me to be obedient to the Father no matter what, and that would grow with me on

this journey with Jesus. Little did I know that social media would be my tree. Read the story of Zacchaeus (Luke 19) to better understand this reference. I arrived at a divine appointment when a mutual friend placed me in communication with Olivia; my olive tree. She is the literal David to my Johnathan, for our souls had been knit together at creation. We are both prophetically gifted and confirmed the Holy Spirit's giftings over each other's lives, but it wasn't until we made contact that some of my own giftings were brought to full fruition. Not shortly after we met, the Lord spoke to me saying that she NEEDED to be a part of writing this book. I went back and forth with the Lord saying, "I can write this myself! I don't need anyone to help me!" God quickly shut me down and gently told me, "You do not know what I am going to do with the both of you, for you need her and she needs you. She is My Paul, you are My Peter. She is My Deborah, you are My Jael. This is only the beginning of how I will use the both of you to glorify My name across nations. Do not reject her, for I have sent her to you to bring this book to completion. I have answered both of your prayers for a friend that is filled with My Spirit and who seeks after My own heart. Listen to Me."

Not shortly after my conversation with the Lord, I spoke to Olivia and told her that I'm writing a book, and that the Lord instructed that was going to be a part of writing and editing the entire book with me. With a smile, she also questioned why God chose her as well, since it wasn't long after we met that we were given such a major Kingdom assignment. This was the assignment the Lord had spoken — I'm sorry, DEMANDED — that we finish together. After we prayed and sought the Lord for confirmation, Olivia began to tell me about a vision that YHWH had given her 3 years ago. As I was listening, I didn't think much of it, but the Holy Spirit started showing me the same vision as she was speaking. I began to give her the details of the plane that she was on, who was on the plane with her, and the destination that she was headed to. Her eyes lit up, she leapt in the spirit and said, "You're the one next to me on the plane and the book in my hand was ours!" Afterwards, we praised and worshipped YHWH for His goodness and for His confirmation of what He had entrusted us to bring to life. We may not always know why we go through difficult trials, or why certain people walk in and out of our lives, but I pray that you never miss a divine appointment that YHWH orchestrates. When God brings something together for His glory, you'll understand that your previous setbacks were just a setup for something greater in store.

    I'm not mentioning any of this to boast in my own glory or to take credit for anything that the Father has done in my life. Everything that I have is because the Creator of the universe has blessed me with it, and I'm painfully aware that He can remove all of it at any given point. This book is solely for the glory of God among His people and for those that would dare mock Him. Heaven's Blueprints is not a "cool" concept with some potential theories. This is the truth from the mouth of YHWH Himself. This is the evidence that even His children have been searching for. Let he who has an eye to see and an ear to hear receive these truths from the Lord. Wake up, saints, there's work to be done for the kingdom of heaven!

OLIVIA

At 5 years old, I was shown hell. This entire nightmare and every second in between is permanently recorded and stored into my long-term memory. You might assume that I, a young child, probably just drew up a horror scene inside of my imagination and ran with it, and I would certainly agree with your point if this had only been based on sight. It wasn't just what I saw that convinced me I wasn't alone in my bedroom, it was what I had felt on that cold and lonely night.

I still remember the pink quilted duvet stitched with white lace flower trimmings laying across my bed. I had pulled it up in order to cover my half-squinted eyes, when I noticed that I almost untucked the bedding from the bottom of the mattress that would leave me exposed to the monster in my room. I was being watched, invasively and intrusively by something that I naively mistook as a harmless shadow. The upper right hand corner of my room started to darken, and to save you the questioning, it was at an hour way past midnight. Unfortunately, I can't tell you the reason why I woke up or what jolted me out of slumber, but I don't think that's an important detail to recall. Maybe it was from sheer exhaustion, or quite possibly crippling fear, that any time I blinked and glanced up at the ceiling, the stretch of the shadow's body was being drawn closer to the center of my bed. I had looked outside to see if something was blowing in the wind, hoping that this looming shadow was just a tree going astray in the wind, but all was quiet and still.

    What I saw wasn't an animal and it wasn't fully human. It was a hybrid of both, morphed into a creature that when examined, my heart rattled inside of my rib cage. If I had the words to describe to you what I had felt in that moment, I would try my best to articulate those emotions, but it was beyond human comprehension. The fear, the booming silence and the echo of unidentifiable screams thundered throughout my bedroom. I had to be asleep. There was no possible way that I was in the same room that I woke up in that previous morning. I peered out from under my covers in an attempt to make eye contact with the uninvited beast lingering in the corner. Its eyes were a dark yellow, its mouth expanded in width and length, and it had 3 rows of teeth that no predator on earth could pair in comparison. Its claws easily passed my entire frame if I stood up straight and with one swipe of its finger, it could rip my flesh in half. I couldn't shut my eyes hard enough. I was starting to sweat profusely and it wasn't due to my duvet almost suffocating me. It was like I was dying without the release of death. I couldn't speak, I couldn't breathe, I couldn't move and whenever I made eye contact with those hollow yellow eyes on that slyly disguised shadow, the first and only thought I had in my mind was, "Nobody is going to save me."

 As you can obviously tell, I survived what I thought was going to be my last night. What I thought was a living nightmare or a bad dream was actually going to be something the Lord revealed as a gift that He gave me. How is seeing hell a gift? Don't worry, I asked Him the same question. Not only did YHWH show me hell, but heaven and the spirit realm here on earth too, each one in full detail. I never thought in a million years that the Lord's purpose for my life was to be an author, but now I know the reason why. God gave me the gift of being a seer (an Old Testament term that we now know as a prophet), so that what I was shown by the Lord in the unseen realms can be relayed in writing to you. This only marks the beginning of where the Lord has taken me.

18

PROLOGUE

"I don't know who's going to write that book, but it's definitely not me," I laughed to myself after the Lord showed me a vision back in 2018. I sent my friend a detailed voice memo relaying everything that I saw and we discussed it back and forth for a bit. I brushed it off and pressed it down because I figured the Lord pressed 'send' to the wrong seer. I won't leave you guessing about what it was that I saw, so I'll tell you.

It looked to be dusk or dawn. I couldn't make out which one it was since I was 30,000 feet in the air and flying first class, which from that alone I knew that this was from YHWH and not a repressed memory. I could see the orange hued tint touching the tips of the clouds as the sun waved hello through the window. I was holding something in my lap and could feel my hand running across the cover of it. A book. Why was I holding a book? If this was my Bible, that would make more sense, but this wasn't a book that I was reading, this was a book that I wrote. "I wrote a book? Where am I going?" I looked to my left and there was an open seat next to me on this flight to a place unknown.

It was occupied, but I wasn't allowed to see who it belonged to. I'm not surprised that the Lord withheld that information from me because I'd be going off of sight trying to find out who it was, rather than walking in faith as He led me towards that person. This wasn't a domestic flight, this was transatlantic. I'm talking more than 20 hours on board a flight, holding a book that I apparently wrote with someone that I didn't know the name of. I landed safely, but I definitely hit major turbulence along the way. I'm a writer and have been an amateur one since I was young, but I leaned more towards writing poetry as I was growing up. I found myself bantering with the Lord after the vision ended. "A book? MY book? OUR book? Is there a pass I can use for this one? I'm supposed to be a nurse! You told me that my purpose was to help people!" The beautiful thing about heeding the voice of the Holy Spirit is that when the Holy Spirit says no, it means no. A firm and unchanging N-O. I'll save you some uninteresting details that happened in the middle of my wrestling, but I did end up writing a book. Not this one, but one that I initially believed was what the Lord showed me in my "this can't be real" vision. The Holy Spirit poured out supernatural revelation and even revealed details that I wasn't allowed to include in that one for the Lord said, "Save these visions for another book that I will give you." The book was completed and I'll leave it at that. I heeded and obeyed the voice of YHWH and I sighed with relief knowing that my assignment for the Father's glory was completed. I could sit back and relax for a minute now that it was out and published. Sike!

    I met Destiny, randomly in my eyes, but it was a divine appointment in the Lord's. I thought YHWH had a sense of humor when she told me that she was going to be writing a book and that I was going to be a part of it. "Lord, I love You and I trust You and all, but didn't You just see what happened during the first one? Can't You send someone else to help her? It can't be me!" Since I'm the one typing right now, I'm sure you guessed His response to my complaints. Little did I know — which is why I'm so glad that I serve an all-knowing God — that this would not only be an answered prayer for a God-fearing friend, but a confirmation that the vision the Lord gave me was coming to pass. You could say that it was destiny that we met.

This book is nothing like I've written before. Most definitely nothing like I have studied before either. I worked in pharmaceuticals, physical rehabilitation, medical offices and was enrolled in nursing school. I thought I had some book knowledge accumulated in my mind, but writing this book and the revelation He revealed was taking what I thought I knew to a whole other level. I'll be honest with you. While reading this book, you'll experience cognitive dissonance and most likely, you'll want to disagree with every single point that I make. It will go against everything that you've been taught, that you've been programmed to believe and that you've persuaded yourself to be true. You might feel angry. You might be amazed. You might even find yourself in prayer before the Lord asking Him to confirm if what we wrote was true or not. Before you start reading, I hope you pray and ask the Lord for the Holy Spirit to pour out supernatural revelation and confirmation to you. These words aren't our own, but directed and given straight from the mouth of YHWH, the God of Israel.

Heaven's Blueprints will challenge every single belief and theology that you have as a follower of Christ and also as a non-believer. Truthfully, it might be easier to digest and receive as a non-believer, which is why I pray you read this book not with your brain, but through the Spirit guiding you. Go into this by leaving your past perspectives at the door and try reading it with an open mind and a teachable spirit. I promise you that this is not the type of book for the faint of heart. YHWH doesn't care if you've been serving in ministry for 50 years, if you've been an Atheist your entire life, or if these findings go against your personal beliefs. To be quite frank, I also don't care if this book offends you because I'd rather stand before God in my obedience in serving Him, than to offend Him by remaining silent in order to cater to people's feelings.

This will challenge everything you think you know, because honestly, you really don't know if you don't have the wisdom that only the Holy Spirit gives. Your human wisdom pales in comparison to the supernatural wisdom of God. Writing this book challenged the both of us to put it down on paper and to lay aside any personal convictions of what people might think about us and also about the Lord's revelation. You might find yourself reading this book because you prayed for answers that you have yet to find, you scorned for "proof," or simply just because you're curious to see what all of this could possibly be about. Set your pride aside, set religious doctrine aside, set your own wisdom aside and read it through the guidance of the Holy Spirit. I'm praying that the Lord convicts and confirms His truth over every pair of eyes that come across these truths from heaven.

"O Timothy, keep that which is committed to thy trust, avoiding profane and vain babblings, and oppositions of science falsely so called: Which some professing have erred concerning the faith. Grace be with thee. Amen."

1 Timothy 6:20-21

Chapter 1:
In The Beginning . . .

"God created man in His own image, in the image of the [Triune] God He created them; male and female He created them."
Genesis 1:27

I remember my Sunday school teacher repetitively reciting Genesis 1:27 right as the clock struck 9 A.M. During my adolescent years, my mornings at church would start off the same and I eventually graduated with Genesis 1:27 permanently ingrained into the center of my hippocampus. Now that I'm a seasoned alumni of River of Life Church in Chattanooga, TN, I can't help but to wonder if any of my past classmates can confidently say that they remember that recited verse and if so, what it could possibly mean through their eyes. An important point to note before I break down this scripture, is that Genesis 1:27 isn't just a reference to mankind's physical appearance. I'm curious as to what your personal reaction would be when you find out that YHWH, the Creator of the universe, created mankind in His own image. Would you smile in awe at the revelation as you notice a glimpse of your reflection in the mirror, or would you merely smirk and push the reality to the back of your mind? Is it humanly possible to try and conceptualize the care and the tenuousness that the Creator had while knitting you together inside of your mother's womb? The entirety of your being, from the inside out, is created exactly in His likeness.

    Like I had mentioned previously, Genesis 1:27 isn't definitively related to the exterior appearance because it's actually rooted all the way down to a person's interior design and their microscopic makeup. The Creator of the universe loves you so much, that He not only created you in such a way to physically resemble Him, but He created the form of man in the likeness of the very plan He had made to save the people of the world; His Son. Have you ever stopped to admire the vast expanse of creation around you? Look at how seamlessly the sun rises above the horizon right as the moon tucks its face away until dusk the next day. Do you notice the effortless transition of how the rays of sunlight are brought forth in order to drown out the darkness of the night sky? Each beam of light that is etched across the sky slowly dilutes the darkness in order to bring you a new day. In the same way, the light of God's Son makes the darkness of evil flee from your path when you stand in His presence. The Father meticulously put the time and effort into His construction of the day and while doing so, He made sure to match the design and the purpose for what Jesus represents in our everyday lives as well. Isn't that insanely beautiful?

  Whenever I decide to stop and acknowledge the Father's creation, I can't help but admire the precision and the attention to detail that was paid to each design. God is, and always has been, a wondrous God. For an example, notice the way an artist creates their portraits. First, they study the blank canvas in front of them by evaluating the amount of space that the frame can hold. Next, they sketch a composite design and begin to let the creative parts of their brain carry the pencil. They contemplate the pigments that they will use in order to create various spectrums of color. Once those hues have been mixed and perfected, the artist will then take an assortment of brushes that are needed in order to bring their portrait to life. Through this art piece, they plan on conveying a part of themselves to the world. These tasks may seem tedious, and to some people they may be pointless, but to the artist they are more than necessary. When all of these steps are finalized and the artist is ready to paint, they realize that this completed canvas will not only become all that it is meant to be, but also that a piece of themselves will live on forever through what they created from scratch.

  Not only did this happen when God created the heavens and the earth, but it's the exact type of care that He took when He formed you and I. Yes, you became a part of God's inner being. He molded you on a blank canvas. He used different brushes that were coated in different colors. He made each person an original one-of-a-kind design, just like our fingerprints can confirm. God gives every single one of His creations their own purpose in order to reach and become the very fulfillment of that intended purpose for their lives. In Psalms 139, David is telling the reader how God foreknew us before we were born and how He created our innermost being. In Chapter 31 of the book of Job, we are told that we are knit together in our mother's womb. What does the process of being knit together look like?

There's a mechanism in our bodies that are known as microtubules. Microtubules are hollow tubes made up of proteins that, when seen under an electron microscope, resemble the appearance of yarn. Each microtubule is made up of three (3) sets of three, then three sets of three within each one, and so on and so forth. We can trace the microtubules and smaller microfilaments, which exist within the microtubules, all the way to the knitting together of our DNA [deoxyribonucleic acid]. DNA is the material that holds our genetic code and instructs each cell in our body on how to function properly on the microscopic scale for the larger being; mankind. Each strand of DNA is encoded with the sequential pattern A [adenine], G [guanine], C [cytosine], and T [thymine]. Each of these amino acids (small proteins) hold a numerical pattern as well: 10 [adenine], 5 [guanine], 6 [cytosine], and 5 [thymine]. In Hebrew, 10-5-6-5 is the numerical value for the name of God; YHWH [yodhehwawheh] {10-Y 5-H 6-W 5-H}. God personally placed His signature within the genetic structure of every person to have ever existed on earth. Jeremiah 31 gives the reader revelation of a prophecy that YHWH would write His laws and His word onto our hearts, so that no matter what the circumstance is, we would inherently know that He is our Creator. If YHWH didn't write His name into our DNA, humans could not be a suitable temple for Him to inhabit because He cannot and will not dwell in a place that doesn't belong to Him, and Him alone.

The entire chapter of Jeremiah 18 reaffirms YHWH as the author and maker of mankind by using the example of, *"Can I not do as the potter does?"* He personally molds each fibrous material to create a perfect piece of His creation. He not only took this care into forming our outer appearance, but He also put in place extra precautions when forming our microscopic interior as well.

Questions

1) What revelations have you received from this chapter?

2) What do you expect to gain from the rest of the book?

NOTES

NOTES

39

Chapter 2: The Beginning

"For the life of the flesh is in the blood: and I have given it to you upon the altar to make an atonement for your souls: for it is the blood that maketh an atonement for the soul."
Leviticus 17:11

Before modern day medicine and the study of anatomy, the Mayans and the Aztecs believed that all blood carries a life giving source. Without the blood, there is no life within the body. I believe that's worth repeating: without blood, there is NO LIFE! Naturally, our blood gives us life and the blood of the Messiah, Jesus Christ, gives us life, spiritually, through salvation. Without the redemptive blood of Jesus flowing through us, there is no means of eternal life in our spirits. Follow with me while I take you on a journey through scripture, and use the Bible paired with scientific evidence to prove that YHWH is the sole beginning, end, and author of all creation.

In our lungs, you'll find carbon dioxide (CO_2), oxygen (O) and other various elements. When we breathe in oxygen, there's a gas exchange in our alveoli when we release, or exhale, carbon dioxide and other waste materials. The biblical parallel is when we breathe in the Spirit of God, there is a spiritual gas exchange in our souls, and sin is expelled from the body. The reason for this is because sin absolutely cannot cohabitate with the Breath of Life.

"And the Lord formed man of the dust of the ground, and breathed into his nostrils the breath of life [the Holy Spirit]; and man became a living soul." Genesis 2:7

Once the oxygen and other nutrients enter the lungs and pass through the alveoli (the center for gas exchange), it is carried into a person's life giving source; the bloodstream. The blood is then filtered back into the heart by the pulmonary vein. Once the blood is inside the left atrium of the heart, the blood then travels into the left ventricle and is then squeezed out through the aorta. From the aorta, the blood then circulates throughout the body by delivering oxygen and nutrients to our different organ systems, structures, and tissues. After the oxygen and nutrients are in their assigned places, CO_2 is picked up by the red blood cells and is carried back into the lungs awaiting expulsion from the body. This is a never ending process that the body does involuntarily - meaning that it doesn't require any thought or excessive action in order to accomplish the task - similarly like how we as Christians are called the repentant sinners. We aren't given this title because Christ didn't save us, but because we all have sinned and fallen short of the glory of God. It's the humility of consistently breathing in the life-giving Spirit of God and expelling the waste of sinful desires from our lives. God our Creator cannot exist where sin is being made manifest due to His holiness and perfection.

"If anyone would come after Me, he must deny himself, and take up his cross daily, and follow Me,"

Luke 9:23. The key word to note here is *daily*. This means that we never finish living our life once we decide to follow and obey YHWH. We must choose to lay down our sins and our own selfish desires everyday, and let the Breath of God and the blood of Christ direct our lives. Through this decision, we allow our lives to be constantly made new according to Romans 12:2, where Paul wrote that we need to be transformed and to renew our minds *daily* by the reading and <u>doing</u> of God's Word. A Christian is transformed into a new creation in Christ in the same way that red blood cells (RBC) are continuously being made new.

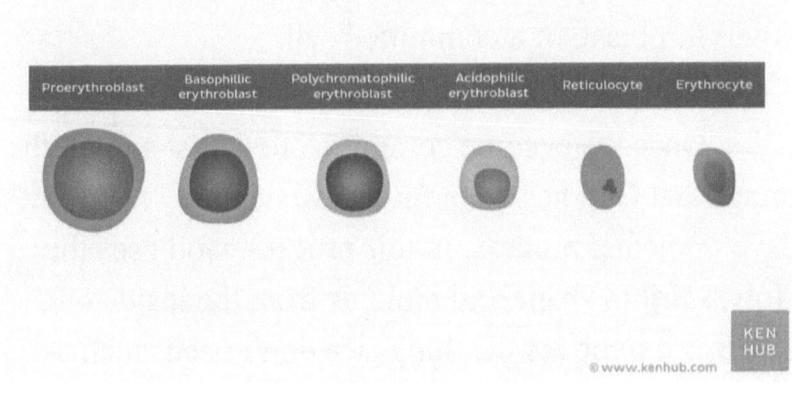

[*Erythropoiesis: Stem Cell->Committed Cell->Late->Normoblast->Young->Red Blood Cell*]

Erythropoiesis is the genesis, or the beginning, of a red blood cell. The process of erythropoiesis begins with a stem cell, which is also known as a hemocytoblast. This stem cell is the natural representation of what takes place in the spirit when we first begin to hear the Word of God or bear witness to it in action. It gains an interest in what is about to take place in its life. The stem cell [you] then faces the choice to either stop and endure apoptosis (cell [spiritual] death due to stagnation), or become something new. 2 Corinthians 5:17 says, "Therefore, if any man be in Christ [saved], he is a new creature." That something new is now what's called a proerythroblast, or a committed cell.

Once we become committed to Christ and all the things that God has done for us, we begin a developmental process. In this process, God uses the Holy Spirit to shape and mold us from the inside out. The Spirit removes the things we don't need, such as: the lies of our past, our sins, toxic relationships, etc., and makes us a new creation. Isaiah 43:18-19 says, "Remember ye not the former things, neither consider the things of old. Behold, I will do a new thing; now it shall spring forth; shall ye not know it. I will make a way in the wilderness, and rivers in the desert." This committed cell will now begin the journey down the developmental pathway.

Phase 1 is the early erythroblast (basophilic) cell. The basophilic cell goes through a ribosome synthesis. Ribosomes are what are needed within the body to make proteins. Proteins make up a person's basic building blocks, which are called amino acids. Without proteins, DNA can't synthesize and a living organism cannot be created. While this is just the beginning, it's an extremely important process. When YHWH begins His work on transforming us, the beginning process may seem long and drawn out, but the usefulness of this process will come in handy later on in the not to distant future.

Phase 2 spawns two (2) new cell creations, the late erythroblast, and then the normoblast. This phase is known as the "hemoglobin accumulation." During this phase, the pigmentation (color) of the cell changes to a pink hue. When the normoblast has attained almost all of its hemoglobin, it ejects or gets rid of most of its organelles [small organ-like structures within the cell]. This is when the normoblast can begin phase three (3), known as the "ejection of the nucleus." The normoblast gets rid of the nucleus, which allows the cell to collapse, and then take its final form. Likewise, when we as believers enter into a place in our walk that feels as if the weight of the world is on our shoulders so much that we may just collapse, and afterwards we have a new outlook (form) and ability to carry on in the rest of our journey. The end of phase 3 is where we can name the current cell a reticulocyte, otherwise known as a young erythrocyte (red blood cell). Similarly, in nature, this current cell can be compared in likeness to a "baby Christian."

At this point, you are making it through your phases, you're being molded, and are now excited about God, but you are still an immature red blood cell. You are passionate about the Lord's purpose for your life and are ready to jump feet first into the deep end, but there's just one thing missing: a lack of wisdom and understanding. There's absolutely nothing wrong with being passionate. YHWH created that feeling and that fire for a reason. A fire that is contained gives warmth and life, but a fire without containment brings destruction and devastation. On the other side of that coin, we can see the more mature Christians who are stuck in the wisdom of their experience, and have left behind the passion of their youth. You need to have both in order to be successful in the purpose that God has placed on your life. Passion without Wisdom is reckless; Wisdom without Passion is lackluster!

You're required to have both the wisdom, and also the passion for your appointed calling in order to live out the experiential life that Jehovah has intentionally ordained for you:

"For Thou hast possessed my reins: Thou hast covered my mother's womb. I will praise thee; For I am fearfully and wonderfully made: marvelous are Thy works; and that my soul knoweth right well. My substance was not hidden from Thee, when I was in secret, and curiosity wrought in the lowest places of the earth. Thine eyes did see my substance, yet being unperfect; and in Thy book all my members were

written, which in continuance were fashioned, when as yet there were none of them" Psalm 139:16-19.

    God wrote that He created you despite your imperfections. He witnessed you in dark places, and He beckoned you out of them. He continually fashions all of those who will be placed onto your new path before you ever become committed. Lastly, the newly formed young red blood cell becomes the erythrocyte that will now carry the oxygen throughout the body: Passion meets Wisdom and lives in the experience of purpose!

Questions

1) Name a time(s) in your life where you felt like the young red blood cell.

2) Does this chapter give you a new outlook on what it means to be a new creation? Why or Why not?

NOTES

NOTES

Chapter 3: Something New

"We were buried therefore with Him by baptism into death, in order that, just as Christ was raised from the dead by the glory of the Father, we too might walk in newness of life."
Romans 6:4

Congratulations, you are now a brand new red blood cell! From the textbook *Human Anatomy and Physiology: 8th Edition* by Elaine N. Marieb and Katja Hoehn, red blood cells are referred to as formed elements. You are a newly created element in the body of Christ [the Church]. A person's most basic function in this new body is to bring forth a breath of fresh air into the other moving parts. Before you feel discouraged and start telling yourself lies like, "That means I'm not good enough," or "I thought I had greatness in me," or "this is all I can do." Remember, without this function taking place in our bodies, none of us would be able to live long enough to achieve the purpose that we are meant to fulfill. To summarize, without your existence, a part of the body wouldn't be able to complete its job. You matter more than you would allow yourself to believe, because now that you're washed by the blood of Yeshua, who is our only means of salvation, you're now a vital member in the body of Christ.

1 Peter 1:2 says, "You were chosen to the purpose of God the Father and were made a holy people by His Spirit, to obey Jesus Christ and be purified by His blood [salvation]. May grace and peace be yours in full measure." God the Father chose us before we knew we needed to be chosen. God formed us in His very own likeness, but He also created us in the likeness of what it would take in order to save the world that He created from the corruption of sin. We carry YHWH's blueprint within us daily. Now, these newly formed red blood cells are suspended in plasma. Plasma is made up of 90% water and 10% of other dissolved solutes. These dissolved solutes include nutrients, gases,

hormones, wastes, products of cell activity, ions, and proteins. The plasma presents the secular [not of Christ] world and it is referred to as non-living fluid. The plasma serves a wide variety of functions, but its proteins are not taken up to be used by the cells to be used as fuels or metabolic nutrients. Though, other plasma solutes will be used, such as glucose, fatty acids, and amino acids. Although we all live in this world, only few will choose to become a new creation (cell) in Christ.

John 3:16 reads, "For God so loved the world [the plasma] that He gave His only begotten Son, that whosoever believes in Him [Jesus] should have everlasting life." The key words to focus on are _should_ and _whosoever_. These two words mean that it is a choice: telling us that anybody can, everybody won't, but somebody will. Going deeper past the plasma fluids, you'll find the fibrous proteins and connective tissues. Both of these are working closely together to create what's known as the clotting factor. The ability that our bodies have to clot is an important, and oftentimes neglected job that our bodies are instructed to do. This involuntary bodily function is what makes a paper cut just a paper cut, and not a life-threatening injury. I have always referred to this as the "tossing out of a life raft" phenomenon. When the skin breaks open, just slightly enough to allow bleeding, this initiates an involuntary response called the clotting factor. The clotting factor is where the connective tissues and the fibrous proteins, also known as platelets, bond together in a weaving, raft-like pattern. This process is creating what we all know as a scab.

While at first it looks unsightly, it's important to understand that under the surface of that wound, new skin is being created and stitched together. Likewise, in the body of Christ, the church is the clotting factor for the world. The sin in your life and past traumas have created open wounds in your heart, soul, mind, and spirit. These injuries might be invisible to the naked eye, but these wounds are bleeding underneath. Jesus, the Savior and Head of the Church, initiates the connective tissue ["Fulfill ye my joy, that ye be like minded, having the same love, being of one accord, of one mind." Philippians 2:2] and the fibrous proteins ["I in them, and thou in Me, that they may be made perfect in One; and that the world may know that thou hast sent

Me, and hast loved, as thou hast loved Me." John 17:23], which initiates the healing process. When we are first being clotted off from our wounds, we still look the same. We are not automatically given a new appearance, but may in fact still feel as broken and disheveled as the scab that we have. Something amazing is transforming underneath the surface though; we are being made into a brand new creation. It's a slow process, but the community, or the body of Christ, is stretching and connecting each person that will be placed onto your path to help bring forth new healing. Once the new places begin to take root, the scabs and scars of your past begin to fade away. The hurt doesn't exist anymore, and when you touch where the wound once was, it doesn't burn. It is now just an event that happened, and the newer, stronger skin will not rip open due to the same situation again.

As we move past the clotting process, you will now find yourself as part of a heme group. Heme Groups are responsible for the pigmentation of the blood. This is known as hemoglobin. Each heme group is made up of four (4) polypeptide chains (two alpha chains and two beta chains), each binding a ring-like heme group. The word heme means "blood" and globin receives its name based on its structure. Since it's globular in nature, the globin also appears as spherical. "We don't connect in a line; we connect in a circle," Pastor Michael Ryan Lindon of Living Faith Church. Expounding on this notion, we notice the structure of hemoglobin having:

1. the globin [spherical structure] and four heme groups

2. iron-containing pigment

It's the iron that makes the key difference between what's known as healthy and unhealthy blood. When a person is anemic, that means that they are iron-deficient. So, how do you prevent the body from becoming iron-deficient?

For our own bodies, we would take iron supplements. Easy enough. But, what about when the body of Christ becomes iron-deficient? We also take iron supplements; I'll prove it to you. Proverbs 27:17 states, "As iron sharpens iron, so a man sharpens the countenance of his friend." The church becomes anemic when people [the iron] leave the congregation for anything other than The Father instructing them to do so. The church is the clotting factor, and if the church isn't aware of its member(s) leaving [bleeding out] this causes damage to the body of Christ, just as bleeding out or not having enough blood damages our physical body. Let's assume though, that the clotting factor is doing its job. At this point, you now have a scab covering the wound that you once had. Now that the hole has been patched [saved], it's the job of the white blood cell to check the body for infection.

*** *I want to stop right here to address the body of Christ for a moment. The reason that you clot off and protect the bleeder before you try to clear out the infection is this: if you try to clear out the infection before you have dealt with the wound, then the body* *will bleed out and die. In other words, we are the hospital for humanity; the wounded get "to belong before they know how to behave." Pastor Michael Todd of Transformation Church.* ***

Questions

1) What scabs of your past are you still dealing with?

2) What has the church, if anything, done to help you Through your healing process?

3) What is something you can begin doing to keep the Hurt from continuing on to others from you?

NOTES

NOTES

Chapter 4: Removing Infection

"The Lord will remove from you all sickness; and He will not put on you any of the harmful diseases of Egypt which you have known, but He will lay them on all who hate you."
Deuteronomy 7:15

There are three (3) basic types of white blood cells (WBCs): the natural killer, the neutrophil, and the macrophage. I have learned through all my studies that since God created humans in the image of the God-head, our molecular structures are also found in patterns of three. Think of the white blood cells as the God-head of our immune system. Confused yet? Don't worry, I'll explain my point in just a moment.

Through this writing process so far, I've realized that the person reading this may not be saved, may not believe in God, and may even believe that I'm reaching. I only ask that you have an open mind as you continue to read. Allow yourself to think that there is a small possibility that I may be right with the points I'm making or you can read this whole piece of literature as fiction, science fiction, but I digress. Let's move on. The purpose of white blood cells is to lead the charge of our immune system. Their design is to recognize foreign bodies that don't belong within the body and to remove them. It's similar to the same way that the Holy Spirit recognizes sin and then leads the charge to remove it from our lives.

The Natural Killer cells "police" the body, the bloodstream, and lymph nodes and are known as defensive cells. Often, they are called the "Pitbulls" of the defense system because of their ability to detect and kill cancer cells and virus-infected body cells before the adaptive immune system is activated. Their primary mode of destruction for foreign bodies is by direct contact and inducing apoptosis [cell-death]; much like the way Christ tells us to die to ourselves [fleshly desires and sinful nature]. *"Verily, verily, I say unto you, Except a corn of wheat falls into the ground and dies, it abideth alone: but if it dies, it bringeth forth much fruit."* John 12:24

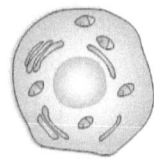 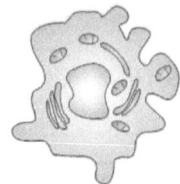 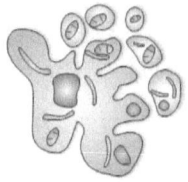

[Process of apoptosis]

If apoptosis doesn't occur, then our body, which is the corn, will die. However, if apoptosis does occur, then the damaged cell dies and the seed of the body (the healthy cells) continue to thrive and will allow for much fruit (life) to be able to spring forth.

The neutrophil is the most abundant type of white blood cell that exists within our bodies. When a neutrophil encounters a foreign body that doesn't belong, it becomes phagocytic. To become phagocytic means to engulf (eat) and digest the foreign, infected body. This is symbolic of how the blood of Christ covers [engulfs] a multitude of sin [foreign, infected body], and redeems the living cell [the human body]. *"But if we walk in the light, as He is in the light, we have fellowship one with another, and the blood of Jesus Christ His Son cleanseth us from all sin."* 1 John 1:7

The third most common type of white blood cell is the macrophage. The macrophage is the "chief" white blood cell. This type of cell is also phagocytic [all consuming]. The macrophage travels in and out of different tissue types, searching for cellular debris that doesn't belong within the organism. The neutrophils are derived from macrophages and are performing the same mechanisms in a more precise way. The same way as Jesus is derived from God the Father, obeying only as God said, and performing miracles only as God said, and in a more precise and personal manner:

> "I can of Mine own self do nothing: as I hear, I judge:
> and My judgment is just; because I seek not Mine
> own will, but the will of the Father which hath sent Me. If I bear witness on Myself, My witness is not true."

> John 5:30-31 KJV

This trio of white blood cells act within our bodies the same as the Triune God-head would. The Holy Spirit detects and removes sin, Jesus redeems and engulfs sin for eternity, and God the Father is in complete control of it all from the very beginning. The deeper we move into the molecular structure of the anatomical design of the human body, the more beautiful this triplicate pattern becomes.

Questions

1) What does apoptosis have in common with Christianity?

2) How can you prevent your own spiritual life from it's own "cell-death?"

NOTES

NOTES

Chapter 5: Energy Repurposed Part One

"The purposes of a person's heart are deep waters, but the one who has insight draws them out"
Proverbs 20:5

No matter what belief system we grew up in, all of us at some point have been taught about a deeper spiritual being that lives within our human flesh suits. I could go on a tangent about everything coming back to one truth, but that may be for another book on a different day. For the sake of this text and my end goal, let's go ahead and agree that the spiritual being that dwells within all of us at some point is the Holy Spirit.

If you've continued this far into the reading, I appreciate your commitment to at least entertain what I'm certain you believe at this moment to be delusions of grandeur. I promise that you are not reading this book by accident. My deepest prayer is that this text reaches those that have either:

    A) felt like they require deeper "evidence of God," and/or
    B) wanted to be able to make deeper connections to God's intelligible design.

    Whatever the reason may be, glory to God that this text has reached your hands to be read by your eyes. To understand what I mean when I make the connection to the Holy Spirit, I now ask that you leave behind any religious preconceived notions and ideologies of who or what you believe the Holy Spirit to be. I pray that you do not allow pride or religion to block you from receiving the revelation. Now, follow me back to Genesis to better understand creation.

Genesis 1:27 reads, "God created humanity in their own image. Male and female He created them." For God to be omnipotent (all-powerful), omniscient (all-knowing), and omnipresent (ever-present), He has to exist outside of time. For God to exist outside of time, He has to be pure energy. We know this because of the law of conservation of energy: *"Energy can neither be created nor destroyed. It IS and can be transformed."* This means energy always was, always is and always will be. YHWH always was, always is and always will be.

In following this understanding of the law of conservation of energy, when God created man in His image, He didn't create new energy. He did, however, manipulate the current existing energy [the God-head] and change that power to form and create all of existence. Recreating man is the simplest part to understand. When God created man, He recreated His own image, like when a painter makes a replica of an already existing masterpiece. So, in what image were women replicated if the God-head is all male? It's easy to think, and in my humble opinion, it's lazy to assume that God just removed the male genitalia and created a female. Try to entertain this thought for a moment, "Or was there an already existing masterpiece waiting to be replicated?" To answer this question, we first need to be in agreement. God the Father was not alone at the beginning of creation and if He wasn't alone, then who was with Him? For the sake of time and argument, I'm going to safely assume that we all agree with and believe in the Holy Trinity: God the Father, God the Son and God the Holy Spirit.

Both the Son (Jesus) and the Holy Spirit were present with God at creation. John 1:10 confirms to us concerning Christ, "He was in the world, and the world was made by Him, and the world knew Him not," and John 1:14, "the Word was made flesh and dwelt among us, full of grace and truth." We can confirm the Holy Spirit's existence at creation with Genesis 1:2, "And the Spirit of God hovered over the waters." The apocrypha [missing text from the canonical scriptures] confirms to us as well that the Holy Spirit was present at creation. The wisdom of Solomon [missing text in the apocrypha; confirmed by Proverbs Chapter 10] Chapter

9:9 states that, "Wisdom was with thee [God]: which knoweth thy works and was present when thou madest the world, and knew what was acceptable in thy sight, and right with thy commandments." This verse also gives us the name of the Holy Spirit: *Wisdom*.

Now that we've confirmed the Trinity was present at creation, whose image did God use to repurpose energy from to form women? Process of elimination confirms that the Holy Spirit is the perfect masterpiece after which women were created. To understand this profound thought deeper in context, follow along with me to Chapter 6.

I know that this is where you really want to stop and call me crazy. Please continue on. We've all been taught to address the Holy Spirit as He. I, myself, am included in that statistic. As trinitarian believers, it can be hard to describe the concept in fullness. If it's 3-in-1 and the Father said, "I am He," then does that make every member He? And if the Son says, "I am He," then it can solidify that notion, but remember, we were created in their image. Not just in likeness, but also in design. Marriage is created as a mirror image of the Trinity — Man, Woman, Child — likewise so is the Trinity — God (Man), Holy Spirit, (Woman) Yeshua (Child). The whole book of Revelation Chapter 12 goes into this in detail, so just stay with me as we continue on. I promise it'll make more sense in a moment.

Questions

1) What have you learned from this text so far?

2) Is this what you were expecting? Better, worse, not sure yet?

NOTES

NOTES

Chapter 6: Energy Repurposed Part Two

"He has saved and called us to a holy life - not because of anything that we have done, but because of His own purpose and grace"
2 Timothy 1:9

    The sole purpose of writing this book is to prove that heaven's blueprints are encoded within us all the way down to the atom, and to accurately be able to do that, we need to examine both male and female anatomy. Within our cellular structures, there is one component that can ONLY be transferred from the mother to her child in the womb, which is called the mitochondria. In 2001, author Bryan Sykes attempted to start and finish the same conversation that we're having right now, but shortly after, he was discredited for falsifying DNA tests that he requested his readers to send in that would determine their "biblical bloodline". Bryan forsook and took advantage of the knowledge given to him by YHWH in order to bring himself glory and unfortunately, his selfish ambitions cost him his reputation and his name.

For the purpose of this text, unlike "Seven Daughters of Eve," we are solely going to be looking at the mitochondrial DNA, not the haplogroups. The reason for this is because the haplogroups better suggest region of origin, rather than familial line. I am not suggesting a familial line either though. The mitochondrial DNA, like an X/Y chromosome from the father, is passed down from only the mother. On a spiritual level, this is why those that are Jewish or of Jewish-decent is based on the mother's origins of birth, and not the father. Mitochondria are deeply complex and they reveal to us numerous revelations about the God-head relationship. Let's take a look, shall we?

On the surface, these organelle (organisms within the cytoplasm of the cell) structures are thread-like, lozenge-shaped, and membranous. They are affectionately known as the power-house of the cell. These organelles are called the powerhouse because they provide most of the necessary ATP supply [Adenosine Triphosphate: molecule that is stored and released as chemical energy for bodily use]. The density of the mitochondria in a certain cell directly reflects that particular cell's energy requirements. Much like how as believers, the amount of giftings (ATP) we have from the Holy Spirit directly reflects the purpose and call that God has envisioned for our lives. Some people may only have the gift of tongues (prayer language) and discernment, while others may have all nine of the giftings of the Holy Spirit. The more mitochondria in any given area is generally a good idea on identifying where the action is taking place within the body. "For where two or three are gathered together in My name, there I am in the midst of them," Matthew 18:20. Even Jesus Himself is telling us that the Spirit hovers above where the action is. The busier an area of the body is, the more mitochondria will be found. But, what I find the most intriguing about mitochondria is their design.

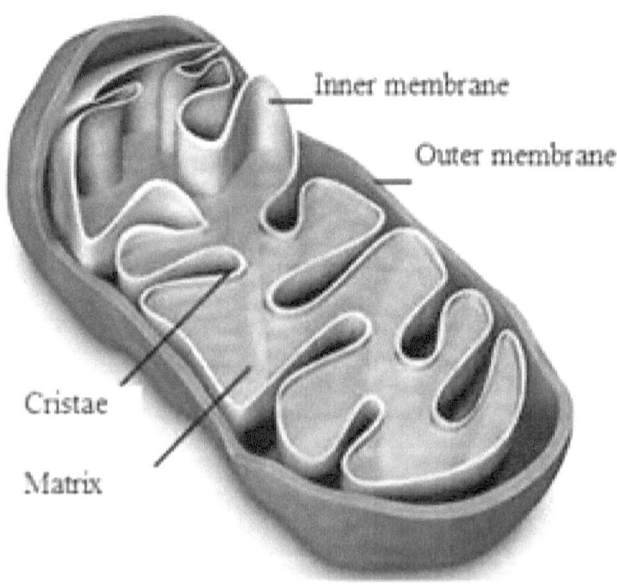

Mitochondria Inner Structure

The mitochondria have three (3) layers. The outer layer, or covering, is smooth and featureless. Much like how God the Father is the outer-covering of the God-head. Smooth in His planning [covering] and how He operates and none on earth have seen His true features [featureless]. The second-membrane exists within the matrix, making many folds and crests to occupy as much space as possible. Likened unto God the Son [Jesus] and how He folds and crests within us, making as much room for the Father as possible. Then, there's the matrix itself. The matrix is responsible for the breakdown of carbon-dioxide (sin) by enzymes by either dissolving it completely, or making it a part of the folded structure. This, I hope you guessed correctly, is the Holy Spirit. The Holy Spirit is responsible for bringing sinners to redemption and making them a part of the folded structure [the body of Christ], or destroying them completely, should that person choose not to become a part of the body. Mitochondria is a natural representation of how the God-head [3-in-1] work together to bring us back into a relationship with God the Father.

You might be asking, "How does this help us understand your first point about motherhood?" The importance of understanding mitochondrial DNA helps us understand female creation, as much as it helps understand male creation. Remember in Chapter 5 how we established that energy is neither created nor destroyed, it is, however, repurposed. For Adam and Eve to have been made whole biologically, they had to have mitochondrial DNA. This begs the question, "Is God the Holy Spirit also God the Mother? *{Author's note: This is noted for continuation of the point and the question at hand, not to dissuade, cause confusion or bring blasphemy against the Holy Spirit in any shape, form or fashion.}*

The Holy Spirit, the power and wisdom of YHWH, will NOT move when grieved or blasphemed. In fact, Jesus gave us plenty of warning for those that would grieve the Holy Spirit. In our many centuries of referring to the Holy Spirit as 'he,' we have seen fewer and fewer miracles, signs, and wonders [that Jesus Himself said in John 14:12 would happen after His ascension to the Father] taking place within, not only the church [where faith for the Spirit to move should be at its highest], but within the world as well. These things have seemingly "stopped," which has led to the ideology of *cessationism*, which is a modern day blasphemy of the Holy Spirit. Why? Because when you believe that a move of the Holy Spirit has passed away and is no longer able to take place, then when you or people you have taught witness this move of God, it's referred to as demonic. Calling something of YHWH demonic is direct blasphemy of the Holy Spirit, and Jesus tells us in Matthew 12:30 that it is an **unforgivable** sin. So, did Jesus lie to us in John 14:12 [negating the word of God because Christ is the Word made flesh], or is there a simpler answer to be examined? We've spent centuries grieving the Holy Spirit, *almost* to the point of an *almost* cessation, not because YHWH stopped being God, but because we as a people have unanimously grieved and suppressed the Holy Spirit.

We are born with the very blueprints of heaven under our skin, yet we spend way too much time debating doctrine and questioning the validity of the Holy Spirit. Effectively, we've unequivocally placed God and all that He is inside of a nice, neat, and easily explainable box by questioning not only if He even *can* still do those things through us, but also by telling others that the Holy Spirit no longer 1) moves, 2) works, 3) gives believers the power to do so, and 4) making it '*impossible*' for the Holy Spirit to be female. Why is that? For some it's just because YHWH is male. How arrogant are humans to believe that an all-powerful, ever-present and all-knowing being that exists outside of time can't be or do anything He needs to reach His creation?

    The proof is inside of our mitochondrial DNA. The same mitochondrial DNA that Jesus was born with. Jesus was born of a virgin into a world of sin, and He was born sin-free. For this to be scripturally sound and for Him to be a whole person — He had to have a sinless, all-powerful, biological mother. God and the Holy Spirit ***WITHIN*** Mary conceived Jesus [not to be confused with God AND Mary]. The key word to note here is **within**. Mary is NOT the mother of God nor is she sinless, blameless or equal to God in any way. The book of Genesis makes it very clear that male and male can't conceive and reproduce children, which leads to the unequivocal response that the Holy Spirit is in fact, female. If God and the Holy Spirit were both male, then there's no explainable reason as to why God would call homosexuality a sin before His eyes because He would've conceived Jesus with a male. That thought in and of itself is blasphemous. It insinuates that God is a hypocrite and a liar, which Scripture says is not true because God is not a man that He should lie.

Before you throw this book across the room and never pick it up again, ask yourself this question, "What did Jesus call the Holy Spirit?" Jesus, God the Son, only ever said these phrases pertaining to the Holy Spirit: "Holy Spirit" "Spirit of God" "Comforter" "Protector" or "My Spirit". The lack of deeper faith and understanding by His disciples, who were mostly men, took it and ran with calling the Holy Spirit 'he' after the fact. So, why would God allow that to be written? The same reason why He allowed the removal of books, the changes of scripture and the mistranslations within the Bible: mankind's freewill. This is why Romans 3:4 says, "Let God be true and everyone else a liar" which does not exempt Paul or the other disciples from this fact either.

Not only does our own anatomy show us these truths, but so many of the missing texts that were removed from scripture. While the "Dead Sea Scrolls" may not be entirely accurate, and as some would argue, completely falsified, their discovery revealed an answer to a question asked by many theologians for the last few centuries: "Do we have the complete divine scripture?" For this book's purpose, I will be using mostly canonical scripture to back my claims, but I'll also be referring to some of the missing texts that I verified myself to be true. Now, you might find yourself questioning how I can even do that since I'm not YHWH. Obviously, you're correct in that statement, for I am no member of the Trinity, only a servant of the Most High. I am also not a messenger angel sent from heaven with a new divine scripture (though if you asked my husband, he'd tell you I'm an angel). A person can verify God's holy scriptures the same way God said them.

When God reveals truth, or a revelation, He does it on multiple levels. Each line of the canonical scripture is identified in at least one, sometimes two, other place(s) throughout the text [refer to Matthew

18:20] and can likewise be verified by 2 Corinthians 13:1, "Out of the mouth of two or three witnesses, let every word be established." Scripture is God breathed, therefore we can also use this information to test the "missing" texts as well. *A faith that has not been tested, is a faith that cannot be trusted.* Therefore, we must ask questions and persevere in our faith in Christ in order to stand boldly before the throne of God the Father.

Questions

1) What does the mitochondria have in common with The Holy Spirit?

2) Have you ever wondered about this same question? if so, did this chapter make the Trinity easier or more difficult fo you to understand?

NOTES

NOTES

Chapter 7: The Bloodborne Substance

"But as many received Him, to them gave He power to become the sons of God, even to them that believe on His name: Which were born, not of blood, nor of the will of the flesh, nor of the will of man, but of God."
John 1:12-13

Now that we have established the recreated (repurposed) energy that led to both male and female creation, how do we further examine heaven's blueprints within our core structure of design? Ninety percent (90%) of the functions within the human body take place in the blood [the circulatory system]. It's extremely fitting that the blood of Christ is what fully saturates us to make us new. 2 Corinthians 5:17, "Therefore if anyone be in Christ, they are a new creation. The old has passed away; Behold! The new has come." This affirms to us that the blood of Christ has to cross our blood-brain barrier.

The blood-brain barrier is a protective mechanism that helps to maintain a stable environment for which the brain can function. The barrier exists because the brain requires constant homeostasis, or a balanced environment. Should the brain in any way lose this balance, it will die and cease to control the body. Hormones and amino-acids that pass through the brain serve as neurotransmitters [nervous system hormones], letting the brain know which parts of the body need assistance and control. This is why Paul told us in Romans 12:2 to renew our minds daily in the Word, so that we as believers may maintain our homeostasis as new creations in Christ.

I'm sure many of you have heard of the Calvinist doctrine of predestination. Again, don't throw this book across the room and never look at this text again, stay with me for a minute. The idea of predestination isn't wrong in and of itself, but how this doctrine portrays itself in the timing of it is. We are predestined for heaven ***AFTER*** we are reborn in Christ. The *after* is an important key word. Still with me? Good! Our anatomy reveals this in the materials known as "bloodborne substances."

Bloodborne substances exist within the capillaries (tiny blood vessels) of the brain. To be able to pass from the brain to the blood and vice versa, these substances must first pass through the three (3) layers of the blood-brain barrier. The first layer is the endothelium of the capillary. This is a thin layer of tissue that allows for entry, likewise how the Holy Spirit leads us to repentance. The second layer is a thick basal lamina surrounding the external face of each capillary. This is symbolic to Christ, for no one comes to the Father except through Him. Now, I want you to hold on to the word lamina because this is important to remember for the next chapter. The third layer of the blood-brain barrier is known as the "bulbous feet" of the astrocytes [cells that assist in the exchange of information between the blood and the neurons]. This is

God, YHWH. We know this because neither the Holy Spirit nor Yeshua [Jesus] move unless YHWH instructs them to do so. The three layers are each acting as one individual, and as a one whole mechanism moving in constant unity and obedience of one, God the Father. Once these stimulants pass through the barrier, they can now move on to what is called pre-committance.

Pre-committance means that they moved to the layers of the barrier knowing that they would enter into the body for a specific purpose and to pass on the information entrusted to them. The neurotransmitters are a small picture of what believers in Christ are to be: pre-committed representatives delivering the information that they were entrusted with, purposed with, and predestined to deliver. Does this sound familiar? If you said, "missionaries" then you're absolutely correct. So how does this apply to the believer not titled as a "missionary?" Matthew 28:19, "Go ye into all the world and preach the Gospel" aka the Good News. Anyone not in Christ is of the world. Yeshua didn't tell His disciples to travel the globe (though it's not a bad thing to do so), He was telling them in the same way He is telling us, to go and preach the Good News to your neighbors. If we are saved by grace through faith and are a part of the body of Christ, then by default we are heaven's neurotransmitters. We are pre-committed to share the information of Christ so that everyone would operate on purpose and in purpose within the body. This is what makes a healthy church structure; everyone shares the information that they're designed to share. Likewise, in a healthy human body, when the neurotransmitters are operating and sharing the information they're supposed to be sharing, the human body is operating in a perfectly smooth fashion. By knowing this information, we can conclude that the holy scriptures of YHWH pass through the blood-brain barrier. By reading and praying daily, it transforms the diseased mind of the sinner into one that is healthy and holy in order to be able to grow into a more Christ-like creation. But, what happens when the neurons [nerve-cells] are misfiring, and the neurotransmitters are sending the wrong information to the body? Is the damage that is caused irreparable?

Most mishaps are stopped by the barrier, however it cannot fight against things that easily diffuse [pass through] plasma membranes such-as: fats, fatty acids, oxygen, carbon-dioxide, and other fat-soluble molecules (bloodborne alcohol, nicotine, and anesthetics). In the body of Christ, this would represent a false-doctrine that is accepted because it makes the believer feel good about themselves. In other words, it's useless fat that needs to be thinned. Then, there are large damages to the brain [central primary theologies that are the foundation of belief in the body] that can increase the difficulty for recovery: traumatic brain injuries, cerebrovascular accidents, and degenerative brain disorders. All of which have different levels of difficulty within themselves [different levels, different devils].

A traumatic brain injury [TBI] usually takes place after a blunt force trauma (a hard hit to the head). When a concussion takes place, it's typically categorized as a localized trauma, meaning that if handled and treated properly, the damage stops in that location and is able to heal in the correct manner.

Though, having more than one or multiple over time can cause cumulative and neurological damage [causing more neurons to misfire]. This is what's known as a contusion. Depending on the location of the TBI, a contusion can be short-lived and the injured individual remains conscious. However, when a more serious trauma takes place (particularly at the base of the brain stem), this situation leads to a coma. A coma is a loss of consciousness that is used by the body in order to take the necessary time it needs to heal, which ranges from a few hours to a lifetime. The individual is still alive, but they're unaware of the amount of damage that their body is attempting to heal. A traumatic blunt force trauma, known as hemorrhaging, happens when the brain is bleeding or swelling. Nearly all cases of hemorrhage are fatal, unless the physician can intervene in time and place you into a medically induced coma in hopes of recovering.

      Cerebrovascular accidents, or strokes, occur when the blood supplying nutrients to the brain has been blocked (usually by a clot), and the brain tissue begins to die. There are two known identifiers for a stroke, a CVA and a TIA. A CVA is a completed stroke, and those that do survive are usually left with at least one side of the body paralyzed. The other is a TIA or a transient ischemic attack. This is called an "incomplete" stroke. While a TIA is not permanent, they are always a "red flag" in indicating that a complete stroke is on its way. A CVA or a TIA doesn't occur when there is only one vein blocked, but rather when multiple veins are blocked and the other vessels are too overworked to pick up the slack from the blockage that is causing other neurons to die or stroke out. What happens though when the damage inside of the body isn't as quick as a TBI or CVA? In hopes to prevent damage, how do we notice the signs?

  Something like this is known as a degenerative brain disorder. The most common of these disorders are: Alzheimer's, Parkinson's and Multiple Sclerosis. While the damage brought on from these disorders cannot be prevented or stopped, they can be stunted or slowed down. The hardest thing to grasp about these disorders is that the early signs are noticed, but they're often overlooked or entirely diagnosed as something different. By the time a person is accurately diagnosed, the body has already begun losing control of itself. You probably have already pieced together how these issues correlate not only with our spiritual beings, but also how they correlate to the church as a whole. If you haven't yet, don't worry, I'll explain it to you.

    When a TBI takes place spiritually or within the church, it's a matter of a sin action that was accepted into the body and ended up causing internal damage that needed healing. This also includes being spiritually asleep. An example of this is the "once saved always saved" doctrine. It sounds really pleasing to the ears, and it makes us feel better about our own decisions, but truth be told, it's entirely false. This invites pride and a superiority complex into the body, which ultimately leads to a body that is spiritually in a coma. There are those that are still asleep, while multitudes have been waking up and are spiritually healthy. This also occurs with those under the guise of the "prosperity gospel." Whole peoples and churches have been placed in a spiritual coma due to their love of money, which is called the spirit of mammon.

A spiritual CVA happens under what I like to call "Deacon and Elder Possessed" congregations. They halt the flow of fresh nutrients (revelations and Holy Spirit giftings) because due to their own pride, they feel like it negates their own. This blockage also keeps the Pastor from being able to receive what YHWH has sent in order to breathe new life into the body. Red flags are noticed when members are being shut down (usually due to appearance), which can be prevented as long as the Pastor (the brain) notices and keeps other pathways open. This allows for the new, or current, members to be able to successfully deliver the information. If no other pathways exist, then the body is paralyzed spiritually and is unable to move at the pace of YHWH's grace for that ministry.

What if it's degenerative? A degenerative disorder can be as small as a difference of opinion within the body. The spread of degeneration can be slowed down if these dissensions are addressed as soon as they arise. However, most address it by simply telling the neuron (the church member) to submit or leave. This behavior being exacerbated is what progresses the disorder to an uncontrollable shut-down of this piece of the body as a whole. All of these, whether as an individual or as a congregation, are what prevent neurons from being able to operate the way that they were designed to operate, which is to go into all the world and to deliver the message of pre-committance [the Gospel]. How can a neuron, a singular member of the body of Christ, faithfully deliver the news that Yeshua wants a relationship with the lost, if the found can't maintain a relationship with each other?

Questions

1) Explain the blood-brain barrier as it relates to the Triune God-head.

2) What is pre-committance?

NOTES

NOTES

Chapter 8: Remember the Laminin

"So we, who are many, are one body in Christ, and individually members of one another."
Romans 12:5

How do we know that the reason we even exist in the first place is to have a relationship with Yeshua? I'm sure this is a question that we've all struggled with during the process of making our human nature line up in obedience with our spiritual existence. If you recall in Chapter 1, we discussed how the Creator [YHWH] signed every single one of His masterpieces with His personal signature. There's also another way that someone is able to discern who the artist of a design is when a signature isn't always evident. It's by the creative style in which the piece itself was created.

There's a molecule within each person, that since its discovery is affectionately named "the God Gene." This is the molecule known as *laminin,* which looks like this:

Laminin is a glycoprotein that acts as connective tissue across EVERY type of cell (tissue, muscle, skin, blood, organ, nerve, etc.) throughout the body. The top three smaller chains connect the cells together, while the larger trunk anchors those cells to the membrane. With this information at hand, I'm reminded of both Hebrews 6:19 which says, "Hope we have as an anchor of the soul," and of Ecclesiastes 4:12, "The one may be overpowered, two can defend themselves, but *a cord of three strands is not easily broken.*" The Alpha Particle, which looks like a "Jesus fish" is representative of Yeshua because He is our anchored hope. It's located at the top because He is the Word of YHWH made flesh, and YHWH set His word above himself. The Gamma Particle [the funny looking "y"] is representative of YHWH because the "Y" in YHWH is Yod/Yah which means "God." Finally, the Beta Particle [the fancy looking "B"] is the Holy Spirit. Why? Not because the Holy Spirit is "second," but because the Holy Spirit only moves at the word and direction of YHWH. A cord of three strands is not *easily* broken.

The reason I emphasize "easily" is because while it is designed to be unbroken, it still can be broken. Our body is a temple for the Holy Spirit to dwell in, but what happens when we desecrate that temple? For the last several decades, we've all heard from the anti-vaxxer community about what a vaccine did to their child, but what if it was actually true and we just weren't paying close enough attention? In 2005, there was a study by a group of military strategists and geneticists about how to stop 'radical' religion. The geneticists tracked the laminin response in the brain of those that were religious and those that weren't and found that those who were religious had a much larger amount than those that weren't. This study lasted for months. They tested different methods until they concluded that there is an enzyme that can both break the bond of laminin molecules and simultaneously aid in the delivery of "synthetic" ones. This enzyme is man-made and is aptly named luciferase. We have only recently heard of this enzyme in the current state of events, but it has been randomly tested in all vaccines since the early 2000s. This is the abomination of desolation, spiritually. Under the new priesthood, which is salvation by Yeshua, we ourselves are the daily living sacrifice unto the will of the Father. When the temple of the Holy Spirit is destroyed, the sacrifice stops.

By being able to understand the blueprints of heaven within our individual bodies and how they were designed, it gives deeper understanding and meaning of the prophecies left for us in scripture. Proverbs 4:7, *"In all your wisdom getting, gain and understanding."* We only know in part and we only prophecy in part, and the reason is not because we aren't allowed to know, but because YHWH desires for us to lean on Him. If we had all of the answers right up front, we would lean on our own knowledge and on our own human wisdom. I find myself adding this in because even the very elect will be deceived. Everyone is searching for something to specifically scream, "I'm the mark of the beast," yet miss the fact that wisdom [the Holy Spirit] knows its number is 666. In tetragrammaton, 6 is the number of man. The mark may very well be physical, but it is just as much spiritual as well. EVERYTHING that happens on Earth has first taken place in the spiritual realm of existence. So, are we relying on Aleph-Tav, or are we relying on ManManMan?

Numerous times throughout scripture we read where both YHWH and Yeshua say that they are the Alpha and the Omega [in the Hebrew language, it is that they are the Aleph-Tav]. I find it evermore fitting that the type of laminin within a certain cell structure is not named for the number of beta or gamma particles, but named for the number of alpha particles that exist within that specific laminin chain. It is a trimeric protein, and remember that YHWH works in threes. This is vital for the maintenance and survival of the tissues because without it, not only would tissues die, but eventually so would the person. Without the redeeming power of Christ's sacrifice on the cross, a person is also spiritually dead.

Laminin is the foundation of our physical being the same way Christ is the foundation of our spiritual being. When YHWH designed us, He designed the purpose of a relationship with us in mind. He also designed us with power and strength in mind. He designed us with Christ's sacrifice in mind. Not that we would only have relationships with other cells (people) within the body, but so that we would also have a relationship with Him. The Father designed this into the core of our being so that in our deepest selves, we would know Him and reach out to connect others to

Him as well. The first thing Yeshua said to Andrew and Simon Peter after telling them to repent and that the kingdom of heaven was at hand, was to follow Him, to connect and be anchored to Him, and that He will make them "fishers of men". What did He imply? That He would teach them to connect and also how to anchor others. Laminin anchors to the membrane first and then searches for rogue cells. It connects first to the rogue cell, then anchors the rogue cell to the membrane, then that newly anchored cell does the same for the next rogue cell. It is continually, each one; teach one. We don't reach the world through mega churches and "fame," we reach the lost by each one teaching one, and creating a cycle of growth and relationships.

    Laminin is active and moving constantly throughout the body by holding it together like glue, while simultaneously reaching out to other cells to hold onto them and anchor them to the membrane. Yeshua is our laminin, and we as believers are called to become Christ-like. This means that we are called to walk the steps of our faith and do the work (obedience to YHWH) that would allow us to reach other cells (people), and anchor them to the membrane, which is a relationship with God the Father.

Questions

1) What is the laminin and why is it important to our Design?

2) How can one, spiritually, be as the laminin?

NOTES

NOTES

Chapter 9: What's My Job?

"For we are His workmanship, created in Christ Jesus for good works, which God prepared beforehand, that we should walk in them."
Ephesians 2:10

This doesn't mean that all Christian's are called to enroll in seminary school and become some type of minister before we can be mature enough spiritually to reach out to the wandering cells, aka lost souls. For this reason, laminin is responsible for the creation of differing maturation pathways. Maturation, or growth of a call happens within a period of three subphases in between a beginning (interphase) and an ending, the G0 phase. Revelation 21:6 and 22:13 read,

> *"I Am the Alpha and the Omega,
> the Beginning and the End."*

The interphase of cell maturation is the period of time between cell creation and cell division. A cell in interphase may be at rest, but it is not stagnant. It is growing within and learning how to operate before it creates another cell. The interphase of a new believer is the time that exists between salvation and spreading the Gospel. A new believer is not stagnant during this time, but rather growing and transforming from within and learning the Word of YHWH and how to replicate (lead others to salvation). The first subphase is called G1 (gap 1). The cell in this subphase is active, synthesizing proteins rapidly and growing exponentially. It is likely that the believer in their G1 phase of Christ-like transformation is learning and receiving their spiritual giftings from the Holy Spirit.

The second subphase is called the S phase. During the S phase, DNA is replicated to ensure that the future cells that form from this original cell will receive identical copies of the genetic material. Without the S phase, there cannot be proper, or correct mitotic phases (where the DNA is replicated). The believer that is in their S phase is gaining and committing to memory the proper information regarding the Gospel of Yeshua HaMashiach (Jesus Christ), and beginning to understand the character of YHWH's heart. Without this information being understood and copied properly and intently, then the new cell that is created, which is a lost soul that is converted to Christ, will be lacking vital information that would make it whole and functional. When a new cell is created with faulty coding, this can be damaging to the body. Why? Because faulty and/or damaged code still continues the replication process, but without the proper coding being passed on to the next cell, eventually an entire system of the body will stop working and die off. This is why the S phase of a new believer in Yeshua is so important.

The final phase of maturation is known as the G2 (gap 2). This phase is also relatively brief by comparison. The proteins and enzymes needed for the replication process are synthesized and are placed in the proper locations. The centrioles (paired cylindrical bodies; each composed of nine triplets of microtubules) are being replicated and the cell is ready to divide. The end of this phase is when DNA begins its replication process:

1.   DNA helices begin unwinding

2.   This is an ATP process [requires the mitochondria]. The enzyme helicase, the Word (Hebrews 4:12) untwists and separates the DNA chain into two complementary chains exposing the nitrogenous bases.

[10 (y) - 5 (h) - 6 (w) - 5(h)]

3. Each chain now serves as a template (set of instructions) for building a new chain.

4. At the site of DNA synthesis, replisomes (mechanisms that carry out replication) catalyze the formation of the RNA primers (10-base long chain) that are responsible for initiating the DNA synthesis.

5. DNA polymerase (genetic glue) aligns the respected bases to their pairs on the new strand of DNA

6. DNA ligase then "welds" the two strands together to form a helix. This is necessary because each helix of DNA includes one old and one new strand because the old is made new.

With all that being said, how would this look on the spiritual level of "replicating" a new believer in Christ? The spiritual process that takes place would look a little something like this:

1. Conviction begins exposing your sins.

2. The Holy Spirit begins dividing what is and what is not of God.

3. This vulnerability now being felt then allows for a person to open up and receive instruction and the knowledge of Christ.

4. The Holy Spirit leads them to repentance.

5. The Holy Spirit begins aligning the person to the gift of salvation through Christ.

6. Now being aligned with the gift, the Holy Spirit seals them into the Lamb's Book of Life and forms them into His new creation.

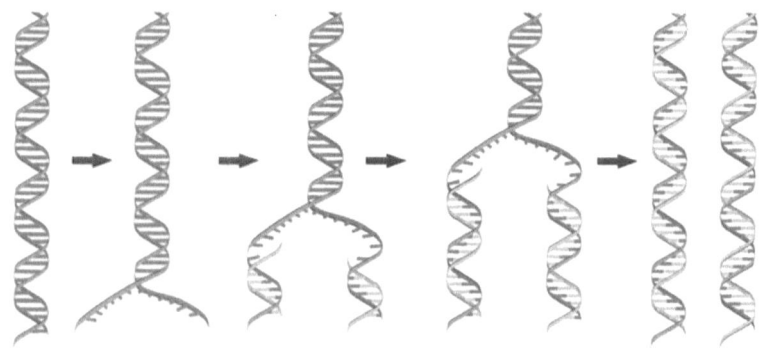

[L to R: Existence; Hearing the Word; Conviction; Salvation; New Creation]

This process can take as long or go as quickly as our humility will allow. Some people begin reaching the lost quickly, while others wait their entire lives to start because of the conviction that their waiting brings them. I am convinced this is why the older generations in the body of Christ frown upon the younger generation. The jealousy of zeal can be so damaging to one's ability to hold to God's instruction, especially when the younger generation is looking to the older ones for help and guidance. Consequently, this is why so many people in the younger cohorts end up leaving the body of Christ altogether.

Questions

1) Why the S phase important as it relates to our walks With Christ?

2) How is the G0 phase bad? How is it good?

NOTES

NOTES

Chapter 10: Stored for Reuse

"Though I walk in the midst of trouble, You will revive me; You will stretch forth your hand against the wrath of my enemies, and Your right hand will save me."
Psalm 138:7

      People from the generations prior that are acting in this manner have permanently ceased dividing and are in their G0 phase. "So let no one despise your youth; instead be an example of the believers, in word, in conversation, in charity [love], in Spirit, in faith, and in purity," 1 Timothy 4:12. When a cell dies unto itself in efforts for the preserving of the body structures as a whole, it's called apoptosis. The G0 phase isn't always a negative thing, though it can be because those members have stopped replicating and are still alive. G0 is also when the cell has fulfilled its purpose for the greater continuation of the body.

      For this purpose we look to Paul who said in 2 Timothy 4:6-7 that it was his time to depart from this earth and be done with his God given assignment. He has run his race with his conviction placed in God, and he reminded Timothy to do just the same. I find it quite intriguing that Paul called our faith journey a race. Why? When we run, our cells absorb as much oxygen as possible to be able to have the endurance to move with us. This is a process known as "blood doping." Blood doping is when oxygen is taken into the bloodstream at extreme amounts in order to be stored as excess fuel for a race. Have you ever wondered how people are able to run and endure such long distances without appearing winded?

      There is a synthetic process that involves doctors and needles concerning some athletes, however this is a process that occurs naturally. If you were to hyperventilate (breath in and out quickly for a very short amount of time), you would feel as if you had more energy and could move at a quicker pace. Pause here and take 20-30 breaths in the next 10 seconds. Your lungs will feel fuller, your eyes are sharper and your body will feel stronger. Why? When oxygen levels are increased, the body has to rapidly produce more red blood cells to be able to transport that oxygen throughout the body. This in turn delivers more nutrients to the tissues, organs and to your brain. It gives you the essence of more energy, which allows you to be able to get more things accomplished. This is what it feels like spiritually when you increase your intake of scripture.

  Matthew 6:33 says, "Seek ye first the kingdom of God and His righteousness and all of these things will be given unto you." What is this scripture truly telling us? In this passage, Yeshua is telling His disciples that we MUST spend our time focused on seeking after YHWH and who He is and what His character is like. We do this by increasing the time we spend in the Word, in prayer and by fulfilling the commands He placed before us to keep us holy, righteous, and set apart. When we devote ourselves to that time daily in getting to know YHWH and His character, we draw closer to Him. The closer we are to Him, the more faith and power we have from the Holy Spirit. The more faith and power flowing through us by the Holy Spirit, the better equipped we are to have patient endurance during this "race of grace."

    Sometimes being a follower of Christ can seem daunting when you're newly transformed or if you're looking in from an outsider's perspective. It can look like "rules" and "regulations," and make you think that Christianity is a salvation of "works." However, it's truly a relationship. Salvation through faith in Christ is not the finish line, it's the starting line! We do not end our faith walk with belief in Christ, we begin it with belief in Christ. Everything that we do afterward is **_because_** of that belief, not in spite of it. A relationship that works, requires work. A runner with impeccable endurance first requires training. There will come a day when we stand before YHWH and we will either hear "well done," or "depart from Me." Yeshua tells us all throughout the Gospels that the world and other believers will know who actually follow Him based on how they love God and love others. The first four commandments in the Old Testament tell us how we are to love God, and the last six tell us how to properly love others. Our natural bodies don't end at life, they begin there. From the beginning, every single piece of our body begins the work it is supposed to do and it doesn't stop until our last breath is breathed. The same goes for the body of Christ. Paul writes in Romans 2:7 that it is by patient continuance that we are made righteous. Jesus tells us in John Chapter 14 how we will continue the work that He began after His heavenly ascension. If God told us that this would be a continuation of existing throughout the church until Christ returns, who are we to stop it? Do we now believe that we are God? We survive to the end with our rewards in eternity, and that should be more important than what people on earth "think" about what God has said.

When the cell has run its race and is entering its G0 phase, what happens next? YHWH calls us the iron of heaven, for iron sharpens a sword [the Word of God], and it is no coincidence that the reason God says we shouldn't forsake the assembling of ourselves is because iron sharpens iron. For a red blood cell, they can't properly function without iron. When a cell dies [it typically lives around 100-120 days], refer to Genesis 6:3, "A man shall not live more than 120 years", so everything but the iron is destroyed. The iron is stored for reuse at a later time. When a believer dies, everything but our soul is destroyed. Our soul is stored [lives] in heaven and is awaiting reuse during the reign of New Jerusalem when YHWH will bring heaven down to a newly formed and sinless earth. The iron stored in the body is held onto until a new set of red blood cells that are needed in order to be created once again. God truly left the blueprints of heaven within our molecular structure, so that in some way, everyone born would inherently feel and know His existence. This is all in hopes that we would all come to Him and be reconciled into a relationship with YHWH through Yeshua HaMashiach and through the conviction of the Holy Spirit.

[Author's note: There are leaflet pages for notes. Go to the epilogue.]

Questions

1) What hope exists for the believer that is considered too young to be leading others?

2) How does being stored for re-use effect the believer?

3) What are your final thoughts of this text?

NOTES

NOTES

Epilogue

You've made it to the end! Be honest, you threw the book at least twice? There were a few times while I was writing this that I threw it myself. Sometimes the wisdom of heaven can be hard to digest, or to even accept the download of information from the Holy Spirit. It's not because we don't desperately want to know the truth, but that we have a hard time making our human existence line up with a righteous God. Which is why we have to die to ourselves, DAILY! God is a God of order. We sometimes believe this means that He operates within the order of things that we know and recognize, but that couldn't be further from the truth. YHWH is a God of His order. What looks like utter chaos to us IS order to God.

YHWH is counter-cultural to the fallen state of humanity. Everything that He is exists in opposition to what we believe in a fallen world. That's why He is simple, but He comes across as difficult or conflated or even cruel. We see YHWH as cruel because His divine truth is offensive to our sinful nature. If we would not do something, we automatically believe that He shouldn't do it. Why? Are we God all of a sudden? Even Christians get it wrong, which is devastating in it's own merit. We are called as believers to be of one mind and of one accord, so that no one is confused by what we preach. Yeshua reminded the Pharisees that no man has seen the Father, but if you have seen Me, you have seen the Father because I and the Father are one.

This wasn't a parlor trick and He wasn't calling Himself God the Father. He is God the Son, but He wasn't speaking to His deity; He was speaking to

His mindset. The design of the Trinity is that they are three distinct persons, with three distinct abilities, working in one accord within their own assignments. Much like the design of marriage and the design of the church. Distinct individuals with their own assignments, but moving in unity both together and apart. YHWH is cyclical. He moves in patterns, not because He has to, but because He wants to be sought out. His patterns are breadcrumbs for those desperate enough to look for Him. So join me and get desperate. I'm time to embrace a birth-right that is given to those that would believe.

Salvation Prayer

For those of you who don't know Jesus Christ as your Lord and Savior and would like to make the best decision of your life, say this prayer out loud, believe it in your heart, and begin the journey of letting the Gospel change you from the inside out.

"Father God, I believe that you sent your only Son Jesus Christ, to die on the cross for my sins and that He rose again. I ask for the forgiveness of my sins and for You to come into my heart. Baptize me with Your Holy Spirit and transform me into Your image. My life is Yours. I pray all of this in Jesus' name, Amen."